Messerschmitt Me 163 Komet

Mano Ziegler

1. DFS 194, flown at Peenemünde-West during the winter of 1939-40.
2. Me 163 A (V3). On October 2, 1941 Heini Dittmar flew this prototype over 1004 kph.
3. Me 163 B-1. This machine flew in combat at Brandis (winter 1944-45). It belonged to JG 400.
4. An Me 163 B-0 (V 28). This test plane was destroyed at Bad Zwischenahn during a landing on August 23, 1944. Its pilot, a member of JG 400, was killed in the crash.

1469 Morstein Road, West Chester, Pennsylvania 19380

Dr. Alexander Lippisch, the father of the Me 163, photographed on the Wasserkuppe (in the Rhön Mountains) in 1929.

Photos:
Mano Ziegler archives
A. Lippisch archives
Profile Publications Ltd. archives

Translated from the German by Dr. Edward Force, Central Connecticut State University.

Copyright © 1990 by Schiffer Publishing.
Library of Congress Catalog Number: 90-60471.

All rights reserved. No part of this work may be reproduced or used in any forms or by any means—graphic, electronic or mechanical, including photocopying or information storage and retrieval systems— without written permission from the copyright holder.

Printed in the United States of America.
ISBN: 0-88740-232-1

This book originally published under the title, *Messerschmitt Me 163 Komet*, by Podzun-Pallas Verlag, 6360 Friedberg 3, © 1977. ISBN: 3-7909-0061-3.

We are interested in hearing from authors with book ideas on related subjects.

Messerschmitt Me 163 Komet

Early Development

The first combat aircraft in the world powered by a rocket, the Me 163 B, could probably also be called the only airplane in whose development nobody could suspect what would become of it later. In addition, it is one of the very few airplanes with a completely false name, for its only connection with the renowned Professor Willy Messerschmitt is that is was built in his factory. Its real father was Dr. Alexander Lippisch. But not even he could imagine during this plane's long period of development what would become of the little tailless bird that he, with so much patience and effort, finally made "fledged."

Alexander Lippisch was anything but a doctor when, in 1920, he first took up the idea of building a tailless sailplane. A few others tried it too but soon gave it up, as there were immense difficulties and bad crashes. Lippisch remained stubborn and, with hard-bitten energy, began to calculate and build model after model. From the few examples that were available to him there was not much to be taken, for in those years between 1921 and 1926 there was only very fragmentary experience in the construction of tailless airplanes. What Alexander Lippisch had to achieve was thus true pioneering work.

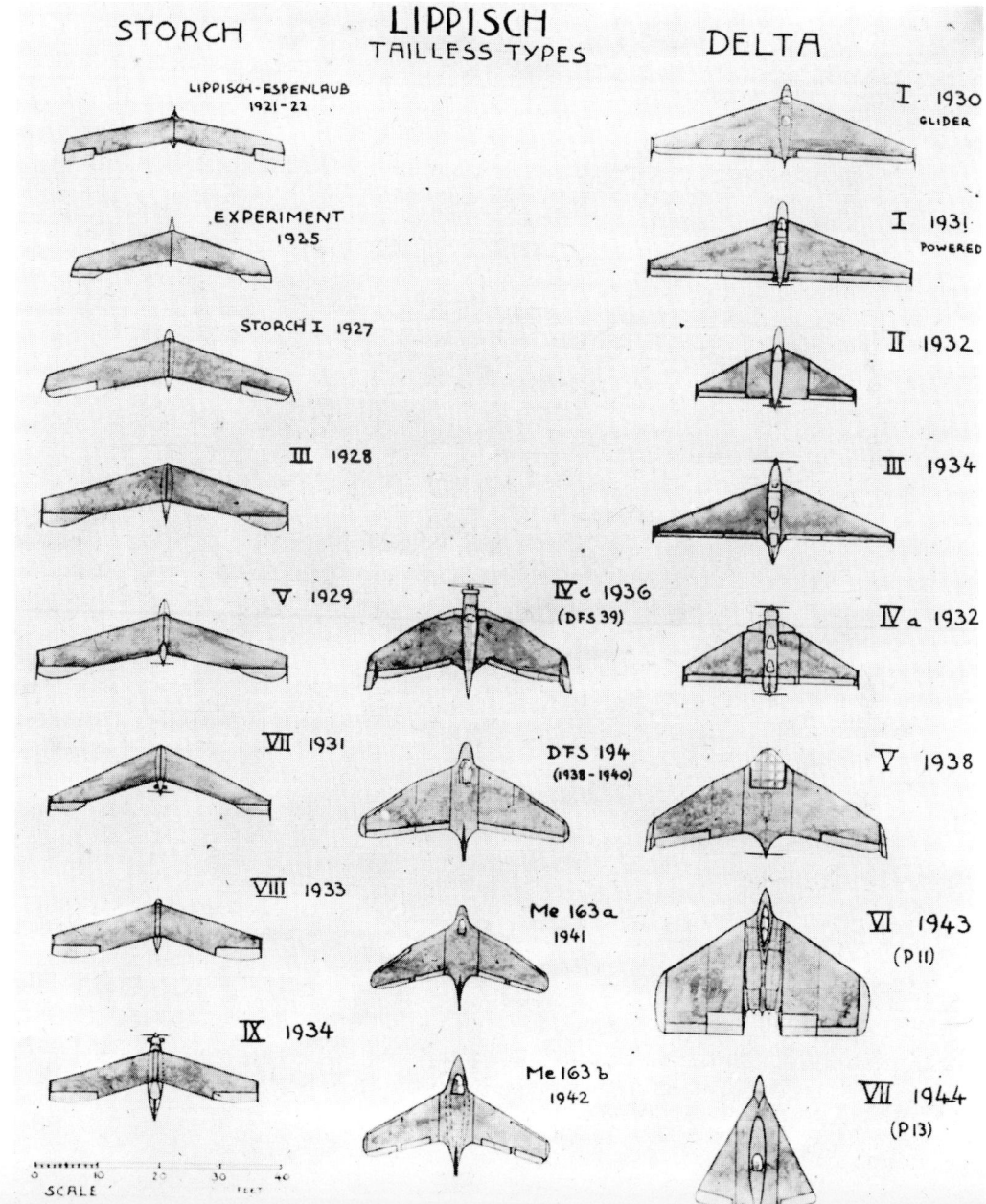

as Aerodynamicist at Dornier 1918

Above:
An early picture of Dr. Alexander Lippisch as an employee at the Dornier works in 1918.

Below and next page: First phases of an Opel-Sander rocket model in 1928.

Only in 1929 could the then world-famous glider pilot Günther Groenhoff take off with Alexander Lippisch's first all-wing glider, which was named "Storch" (Stork), immediately after Lippisch's firstborn son had been baptized with the name "Hangwind."

This event had been preceded by, among others, the first experiments with the flight of rocket-driven tailless models. It was Opel-Sander rockets that powered these models. For Alexander Lippisch, these experiments were not much more than an interesting game that brought him amusement, even when his glider called "Ente" (Duck) was equipped with an Opel-Sander rocket in 1928 and made several successful short flights as the world's first rocket airplane. Despite this success, though, nobody thought seriously about pursuing this type of flying farther, since the rockets only delivered a thrust for a few seconds and could not be considered for longer flights. Still in all, the concept of rocket flight was realized.

In 1928-29 the first "Storch" matured to its variants "Storch IV", a tailless sailplane, and "Storck V", a motor glider with a 2-cylinder 8-horsepower motor, which showed such good flying characteristics with Groenhoff as its pilot that Lippisch decided to present it to government representatives and the press at Tempelhof Airfield in Berlin. After a dramatic flight to Berlin, the presentation there was a complete success, which thrust the young designer Lippisch into the spotlight of publicity at once. It remained so when Groenhoff, while making an exhibition flight in Darmstadt a few months later, was hit by a very strong falling gust at treetop height and smashed to the ground. Miraculously, Geornhoff was unhurt, but the Storch was irreparably ruined.

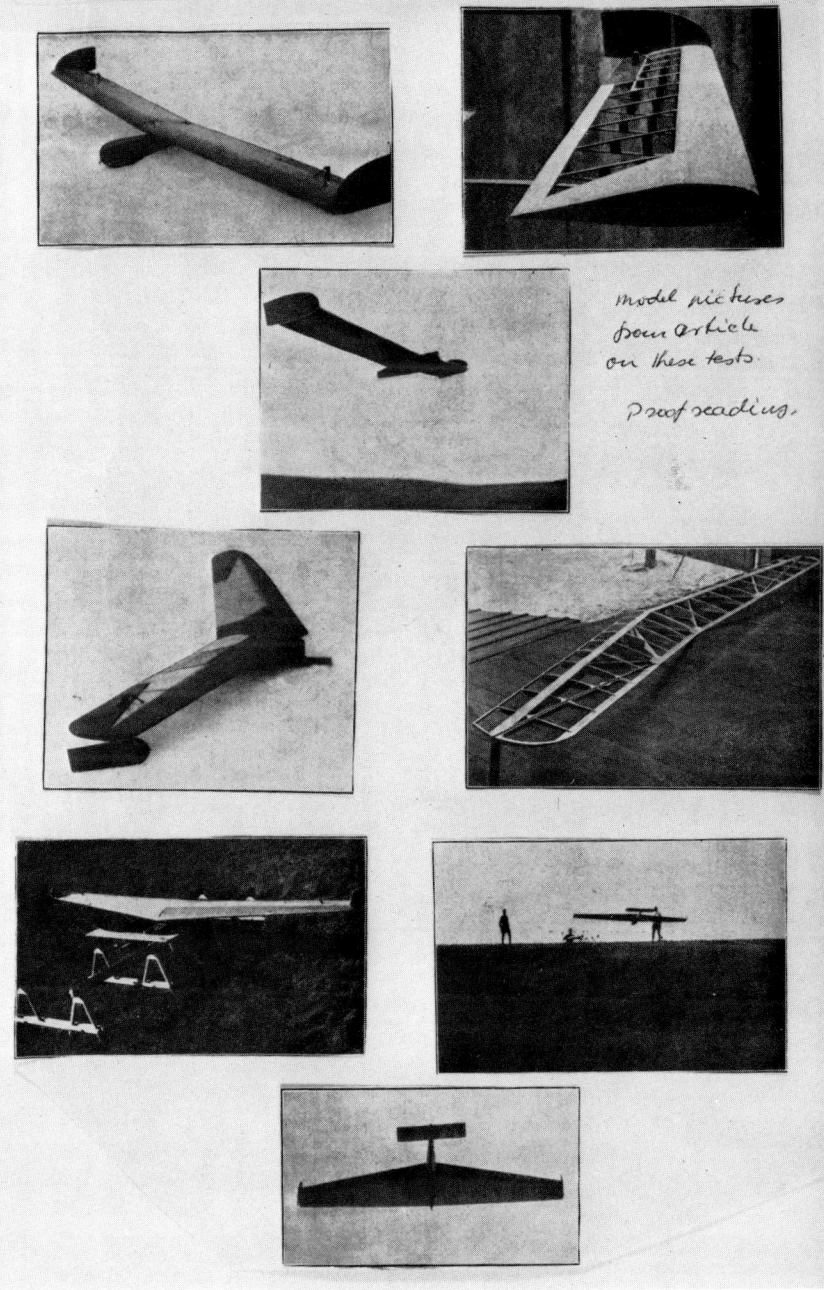

Above: Various test models made by Alexander Lippisch between 1923 and 1928 to test the flying characteristics of an all-wing plane.

Start der Segelmaschine „Ente" Wasserkuppe Rhön 950 m

A rope launch of the "Ente" sailplane (with control surfaces on the side of the fuselage). Particularly noteworthy because Fritz Stamer made the world's first manned rocket-powered takeoff with this Lippisch plane. The powerplant was an Opel-Sander solid-fuel rocket. 1928.

On the Wasserkuppe in 1928 there appeared the remarkably well-flying glider and later powered plane "Storch", with which Lippisch made a real breakthrough in the problem of the all-wing plane. The "Storch IV" shown in the picture is not a delta-wing plane. Storch V was powered by an 8-HP DKW motor.

Lippisch and Günther Groenhoff (in the cockpit) with Delta I in 1931, at the first exhibition at Tempelhof Airport in Berlin.

Right:
During the work of further development of the Delta aircraft, the Storch IX also showed its outstanding flying characteristics at Darmstadt and in overland flights. It had a DKW two-cylinder, two-stroke motor with air cooling.

Günther Groenhoff (at left, next to Lippisch) was one of the most talented and famous glider pilots in the years around 1930, and Lippisch's first test pilot for his all-wing planes, which were often very dangerous in flight testing. Groenhoff, like Heini Dittmar after him, made major contributions to the development of these planes. He died in a sailplane competition at the Wasserkuppe on July 23, 1932. This picture, taken from behind his famous Opel Laubfrosch, is probably the last picture taken of him.

The Delta in flight (1931). From above it seems a bit heavy because of the built-on fuselage with cabin and 30-HP Bristol motor, but seen from below in flight (see opposite page), it shows the striking elegance of its streamlined form, which already makes clear the path to the Me 163.

Lippisch now knew that he was on the right course and began at once to design the motor glider "Delta I", a likewise tailless plane that originally had an 8-horsepower Bristol Cherub motor. With its power, the all-wing plane attained the astonishing speed of 125 kph and showed outstanding flying characteristics. A little later the 8-HP motor was replaced by a 30-HP Bristol Cherub, and the plane was demonstrated to the appropriate government officials at Tempelhof Airport in Berlin. It scored a great success with the press and the public, but the men from the government, though impressed, were not convinced. They had no intention of supporting the young designer's further research.

The Delta I can already be seen as a direct ancestor of the later Me 163, even though the elegance of its form did not match that of the actual 163. But there was time for that. The wing thickness of the Delta I measured three meters at the fuselage, the ailerons and elevators were mounted at the trailing edges of the wings, with the fins and rudders at the wing tips. The cabin and motor were mounted at the center of the delta wing, and the fuselage structure had a pusher propeller at its rear end.

Despite the lack of official interest in Berlin, Lippisch went on building his delta-wing planes, though under the most severe financial limitations. Delta II to IV followed in quick succession, and then a severe blow struck the small construction team. Günther Groenhoff died a flier's death in the 1932 Rhön competition and took all his wealth of all-wing experience with him to the grave.

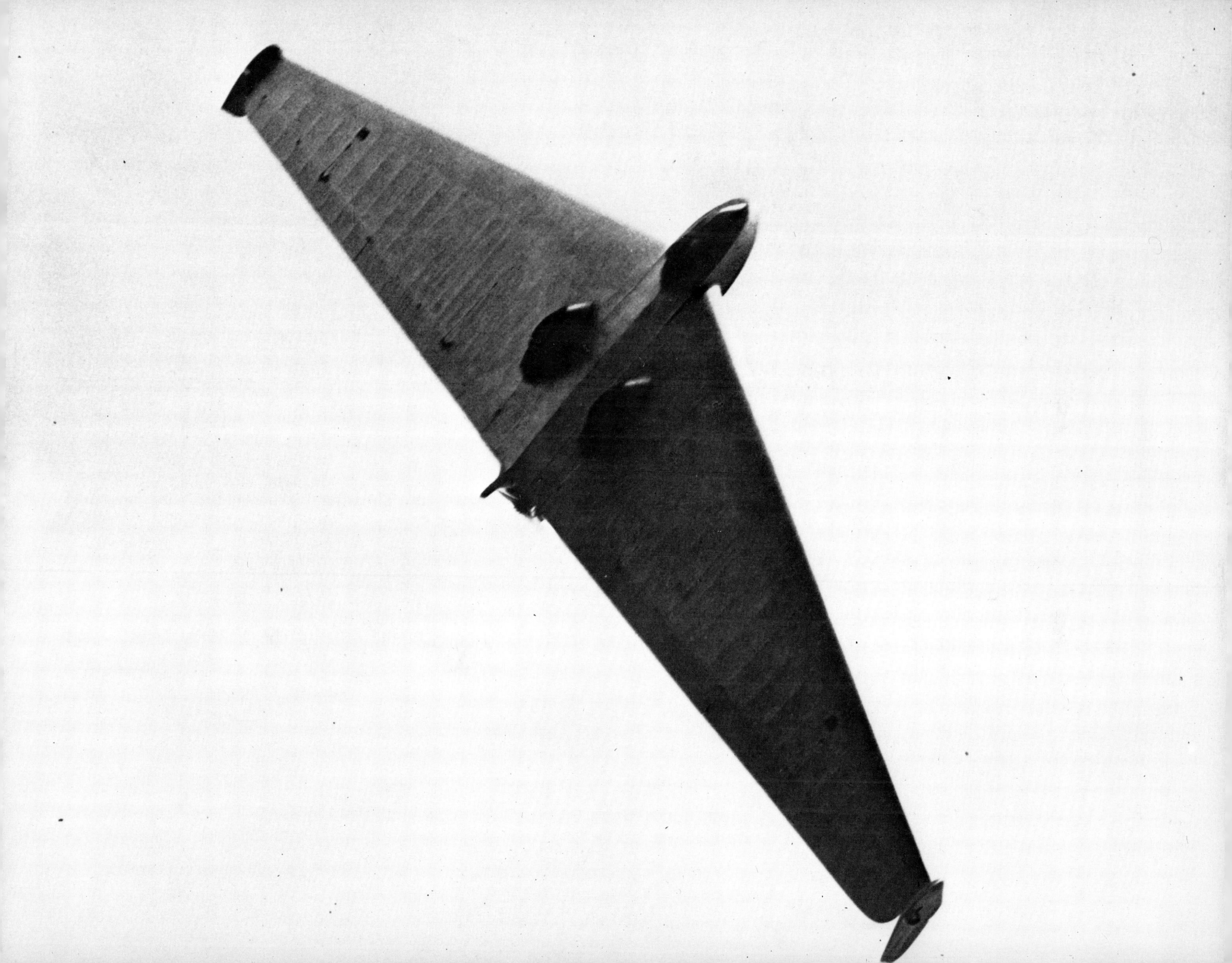

Above: In 1932 this Delta IV appeared, with nose steering, the further development of which did not take place because of unsatisfactory flying characteristics.

Right: Delta I at its second exhibition at Tempelhof Airport in Berlin on October 25, 1941, over the Lufthansa building. In this display, Groenhoff performed some acrobatic maneuvers and an unintentional but safely ended roll.

Opposite page: Lippisch's first model design for a long-range delta airplane, dating from 1932.

Above: A Stork IX during a flight near Darmstadt.

Opposite page:
The DFS 39 with registration D-ENFL was named by Alexander Lippisch himself as the actual forerunner of the Me 163. Heini Dittmar — the successor of Günther Groenhoff — was its talented and successful test pilot and remained so until DFS 39 (originally Delta IV) developed into the Me 163. This D-ENFL was not only a test model but also a good, reliable plane for courier and overland flights. In the Lippisch group it was therefore nicknamed "Our Airliner."

Other setbacks followed. The airplane designer and acrobatic ace Fieseler crashed his own "Wespe" (Wasp), made to Lippisch's Delta IVb specifications, and became an opponent of such planes from that moment on. Groenhoff's successor as the Delta test pilot, Wiegmeyer, crashed in a maneuver at Darmstadt, surviving the heavy crash unscathed. A Delta III built by Focke-Wulf was judged negatively by a Dr. Kupper when tested at Rechlin, and finally a commission of technicians from the Reich Air Ministry and the German Testing Agency for Air Travel banned the further construction of Delta planes. Wiegmeyer and another test pilot, named Tönnes, were killed in test flights.

Now everything seemed lost, but no less a party than the Director of the DFS (German Research Institute for Glider Flight), Professor Dr. Walter Georgii, spoke so vigorously in favor of Lippisch's research that the commission's verdict was not only set aside, but a sum of 10,000 Reichsmark was made available by the RLM for further development somewhat later.

Out of the wreck of the Delta IVb "Wespe" there now arose the DFS 39, to whose construction the aerodynamicist Frithjof Ursinus made valuable contributions. At this time the glider pilot Heini Dittmar, one of the most successful Wasserkuppe fliers, became Lippisch's final and decisive test pilot.

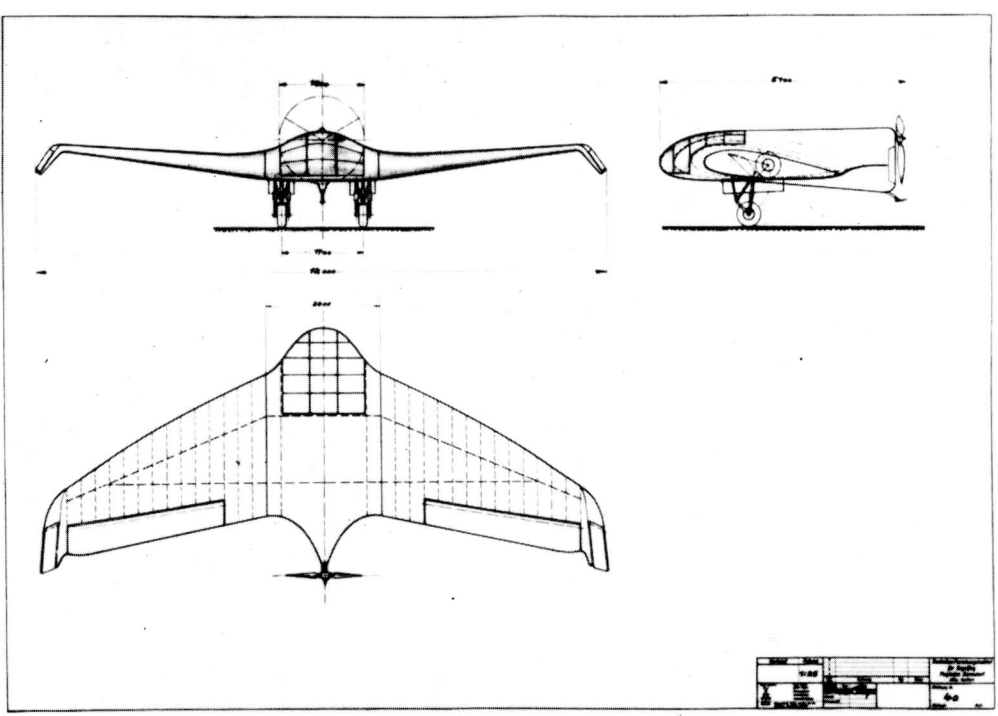

Above and left: In 1938 Lippisch built the DFS 40 as an all-wing two-seater at the German Research Institute for Glider Flight in Darmstadt, but it was not followed up because of more urgent work that began only slowly. The pictures show a three-way design and the craft on the ground.

Opposite and following pages: The pusher-propeller middle-deck DFS 194 was originally built by Lippisch for certain flight research, but in 1937 a very secret contract came from the Air Ministry calling for the installation of rocket drive in the DFS 39. In order to have a second rocket-powered machine for research purposes, Lippisch also rebuilt the DFS 194 (pictures) as a rocket plane, and it took on a special importance for further development when it attained very informative flight data (with Heini Dittmar at the controls) and valuable experience.

The photos of its takeoff and flight were taken at Peenemünde in 1940.

The Secret World Record

The DFS 39 was powered by a 75-HP Pobjoy motor, went into testing with it and kept it without limitations as a two-seat sport plane. This was at first another very good success for Alexander Lippisch, and particularly for his colleague Fritz Krämer, for now they had more or less achieved and proved what they had wanted to achieve and prove: namely the unqualified practicality of an all-wing airplane.

Then in 1937 news came at the arrival of which Lippisch, almost horrified, exclaimed: "For God's sake!" The research office of the Reich Air Ministry gave a contract for a second prototype of the DFS 39 that was to have a slightly changed fuselage to allow the installation of a "special powerplant." This special powerplant was a liquid-fuel rocket which was being built at Kiel by Hellmuth Walter. A powerplant with 750 kp of thrust, a so-called "cold" powerplant with an exhaust-gas temperature of 800 degrees Celsius. (The later powerplant of the Me 163 B had an exhaust temperature of 2000 degrees.)

With the necessary decision to carry out the further development of the DFS 194, as the second prototype of the DFS 39 was designated, at the Messerschmitt works, the construction of the world's first combat-ready rocket plane entered its final phase.

At the Messerschmitt works in Augsburg, Lippisch and his development engineers were assigned several rooms in which the strictly secret "Department L" continued its work. And here the rocket project received its ultimate name of "Me 163." This name was a deliberately chosen disguise, for "Me 163" was the former designation for the development and construction of a short-takeoff slow-speed plane that had been planned to compete with the later-famous "Fieseler Storch." Since the Fieseler Storch had won the competition, the Me 163 designation had become vacant and had the advantage of nobody at the factory knowing that it now referred to a high-speed flight project. All this took place in 1939, shortly before the war broke out.

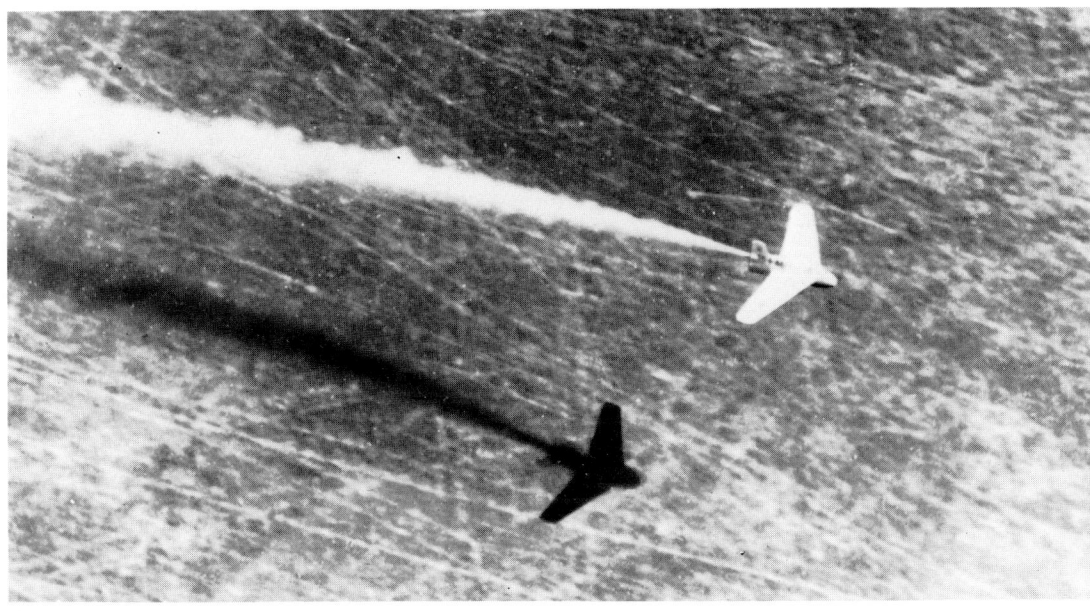

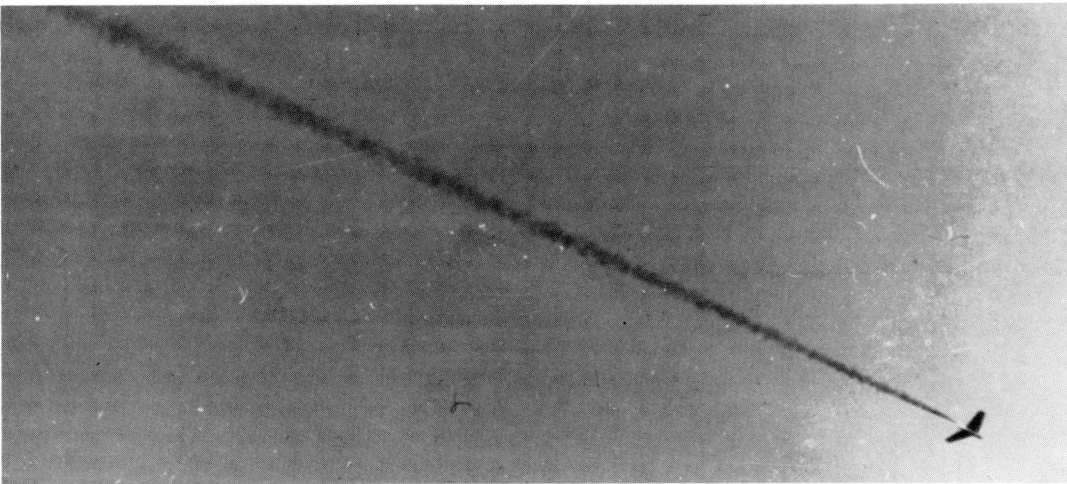

Opposite page and above: DFS 194 during and after takeoff.

Below: A photo of the famous fast-flight experiment in which Heini Dittmar, for the first time in the world, crossed the 1000-kph barrier in the first Me 163 on October 2, 1941.

The test flights with DFS 194, which had been rebuilt into Me 163 A, began in the spring of 1941 with the first towed flights behind an Me 110. Its flying characteristics were outstanding, its astonishing angle of glide measured 1:20! But in its first high-speed flights there were also serious difficulties at 800 to 900 kph. Often the powerplant simply shut off, even at 500 kph. Dangerous aileron and rudder vibration, or even more serious control disturbances, caused great problems again and again, putting the pilot and plane in very dangerous situations. There was also the catastrophic situation that, in takeoffs with the powerplant from an airfield, the fuel was almost used up when Heini Dittmar reached the necessary safe altitude and wanted to exceed 900 kph.

According to calculations, the Me 163 A had to be able to attain the theoretical threshold of 1000 kph at an altitude of 4000 meters if all was to go well. Thus Heini Dittmar decided to dispense with the fuel-consuming power takeoff from the ground and have the plane, with its tanks full, towed to the 4000-meter altitude.

This procedure succeeded on October 2, 1941. Dittmar cast off the towline at 4000 meters, turned on the motor and flew over the measured course, which was equipped with Askania film theodolites. He had already reached 980 kph before the measured course began. Then he saw that the airspeed indicator was over 1000 kph, but shortly thereafter, the elevator began to vibrate and at the same moment the plane, accelerating strongly, went into an almost vertical dive and no longer reacted to the controls. He turned the powerplant off immediately, thought for a few seconds he would have to bail out, but was then able to pull the Me 163 out of its dive and land in a quiet glide.

The first Me 163 A had been completed by the spring of 1941 to the point that Heini Dittmar could take off on the first test flights after being towed behind an Me 110. It had a sensational glide angle of 1:20 and excellent flying characteristics. In the autumn of 1941 the Walter rocket was installed, with which the plane, with the registration KE + SW, soon took off on its first high-speed flights (see previous page).

The first Me 183 B was towed out of the hangar at Augsburg for its first flight tests after a design and building time of only five months. This VD-EL was the prototype of the later combat planes.

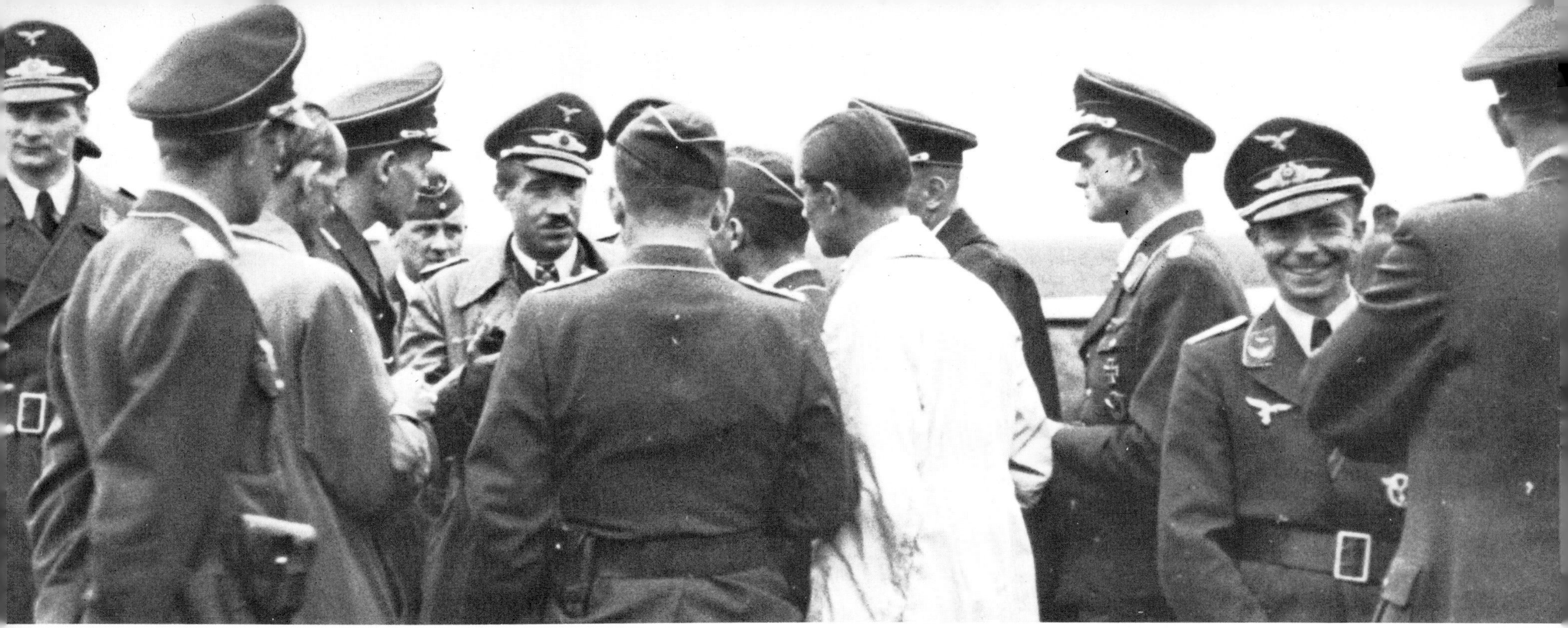

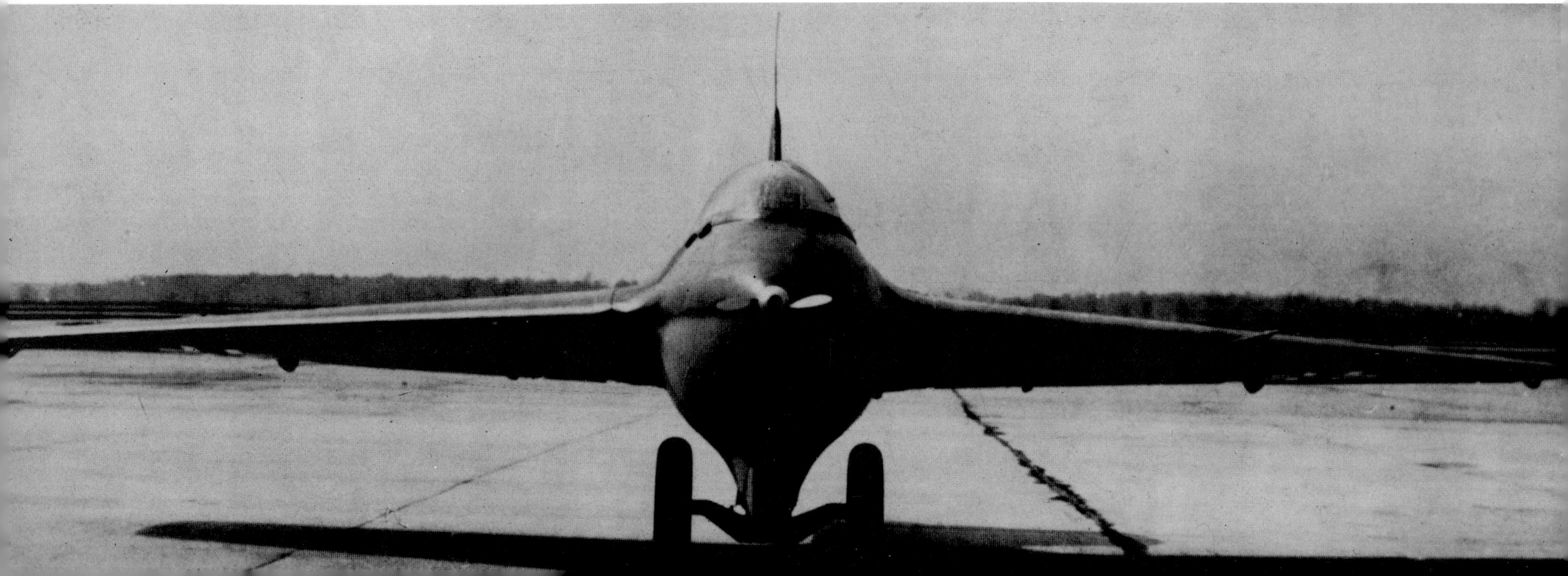

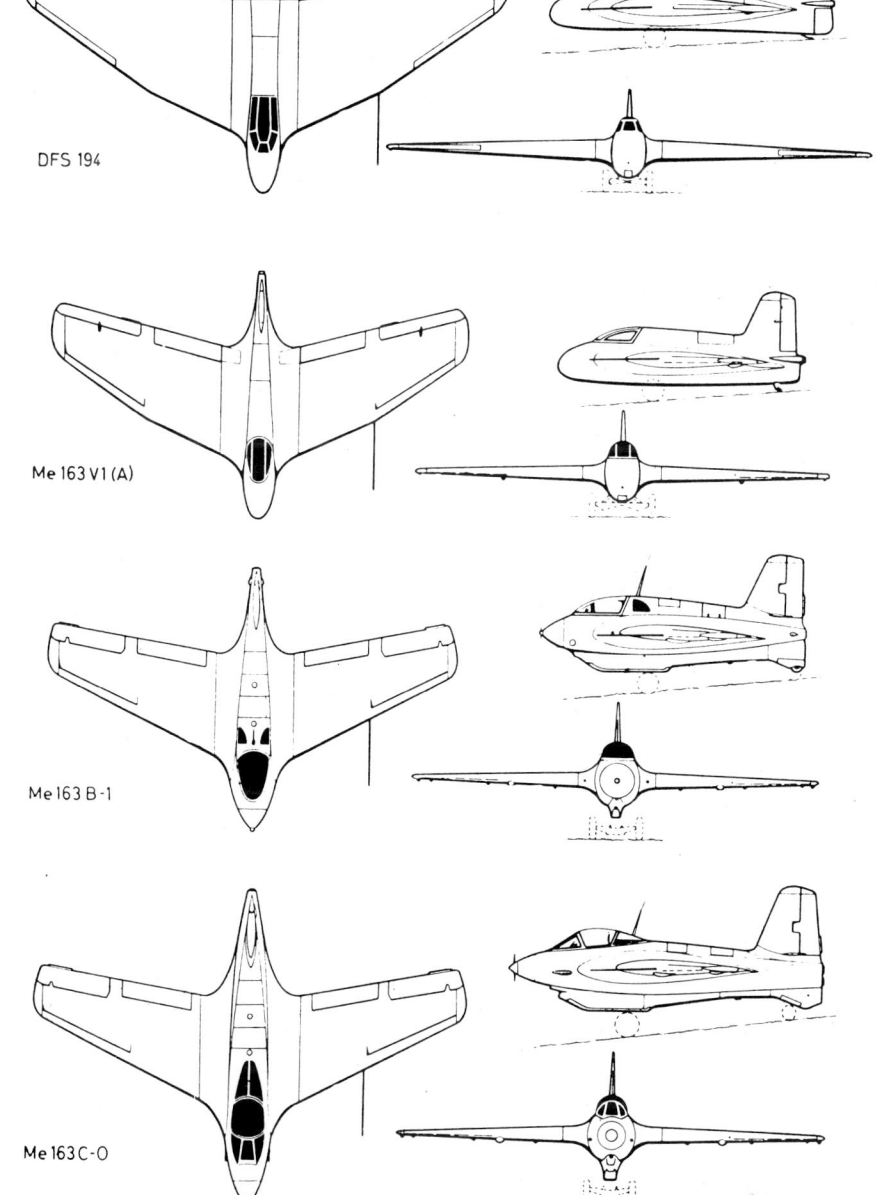

Opposite page, above: The General of the Fighter Pilots, Adolf Galland, attends a preview of the Me 163 A at the Lechfeld in 1942. In white coveralls in Heini Dittmar. To his right are the subsequent commander of "Test Command 16", Hauptmann Späte (in profile) and Oberleutnant Rudolf Opitz, after Dittmar one of the most successful test pilots.

Opposite page, below: An Me 193 B-1b, front view and, above on this page, diagonally from the front.

Right: The development from the DFS 194 to the Me 163 C-0.

Above: The only Me 163 B, still to be seen in Germany today. A contribution of the Royal Air Force made from captured components. The picture shows the plane after its restoration by MBB in Manching. It is now in the German Museum in Munich.

Opposite page: To the front by train. This was necessary because the running time of the Walter rockets was only about five minutes and it was thus impossible to fly the Me 193 B to the Bad Zwischenahn testing site or to one of the later combat bases under its own power. Now and then overflights, such as over Rechlin, were carried out, with the plane towed by an Me 110.

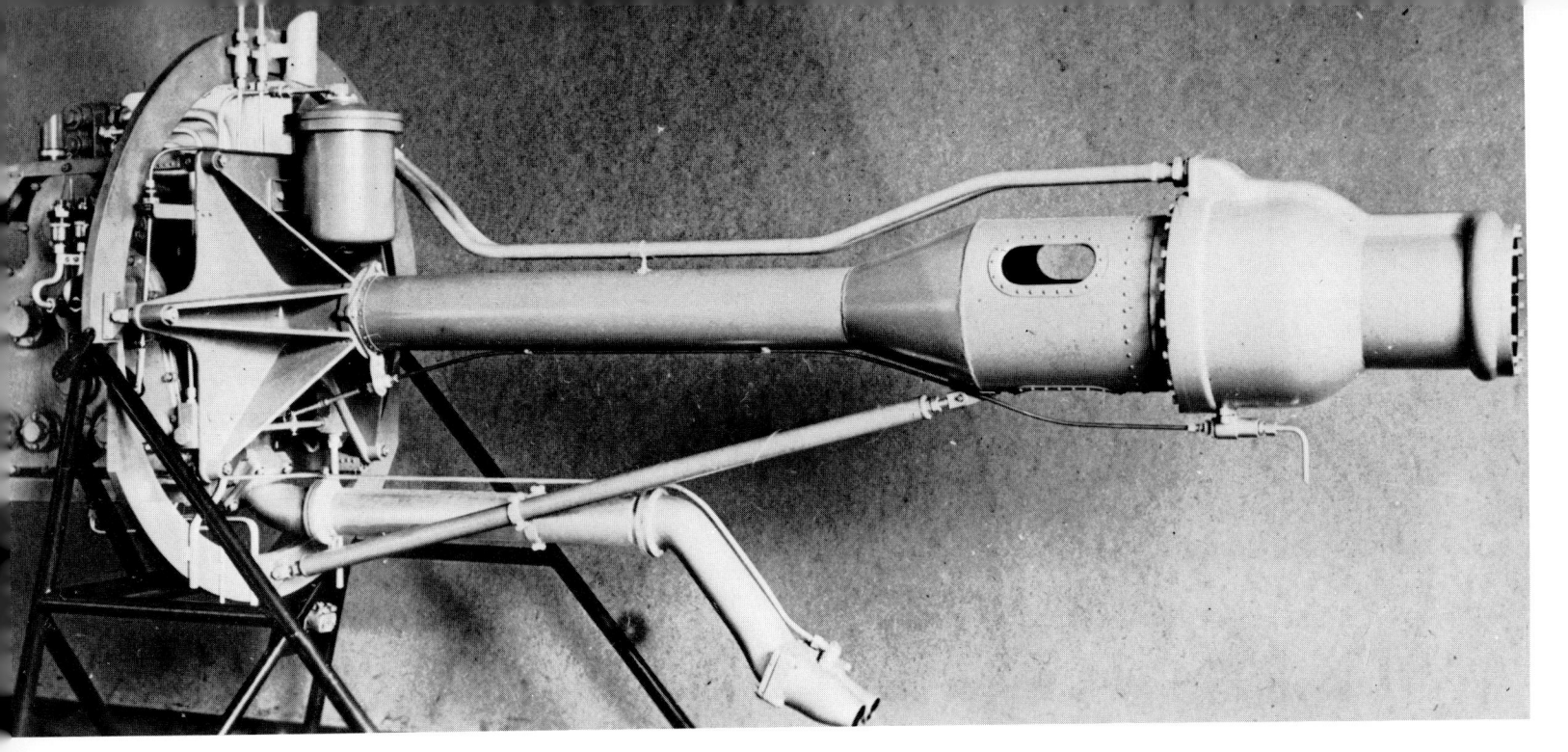

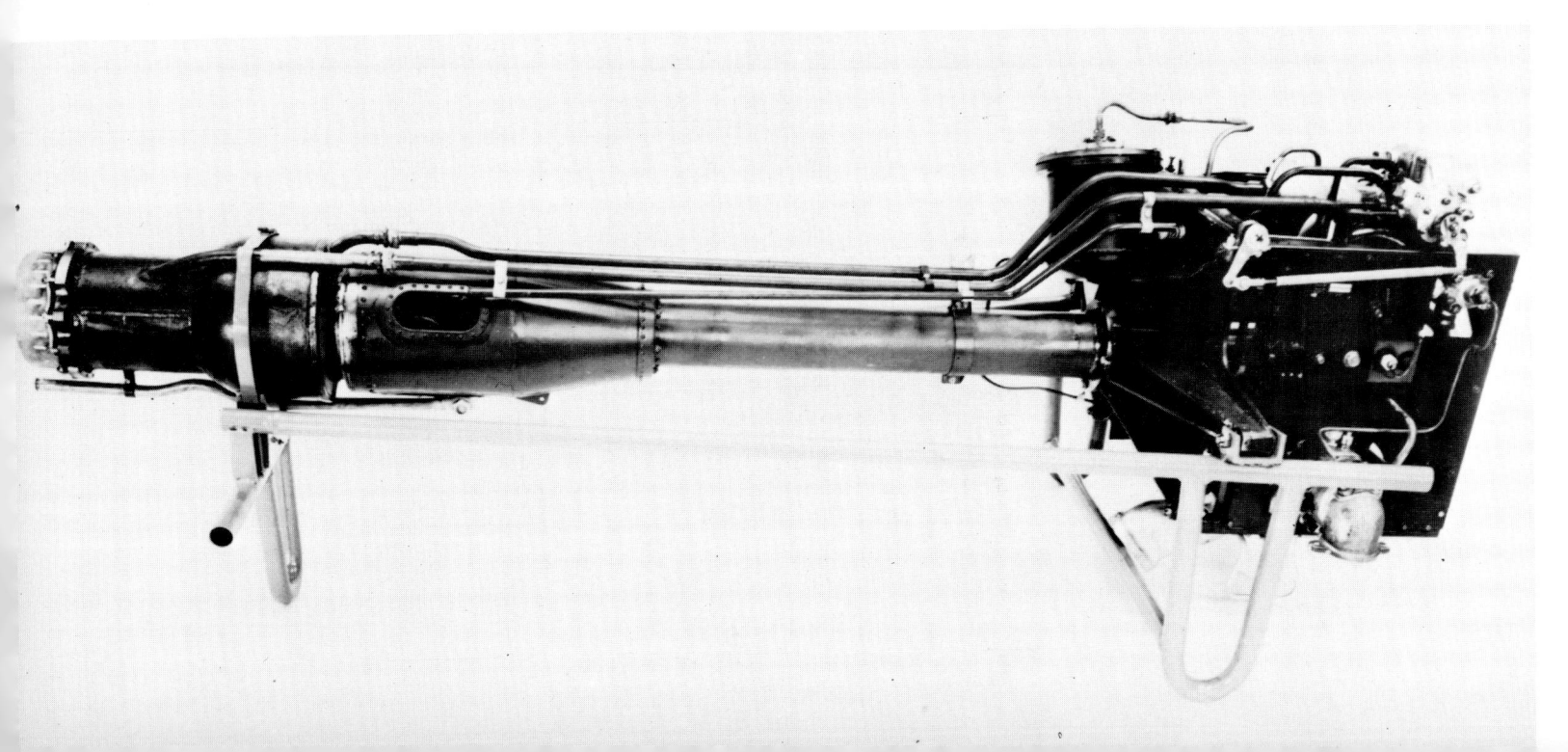

Left, below and opposite: The powerplant of the Me 163 B, designed and built in Kiel by Hellmuth Walter. Its type designation was HWK 109509 A. Controlled thrust to 1600 kp. Dimensions and data: Length 2.532 meters, width 0.900 m, height 0.732 m, volume of burning chambers 9.00 liters, minimum diameter 0.83 cm, exhaust diameter 16.4 cm, dry weight 170 kg, gross weight 180 kg, consumption ca. 2000 kg C- and T-fuel in ca. 5 minutes, average rate of climb 800 kph, top speed in horizontal flight not attainable on account of the sound barrier.

The Askania theodolites had measured the world-record speed of 1004 kph, but like the plane itself, this was so secret that it was never reported. But the "secret world champion" and Alexander Lippisch were delighted, all the more so because the pilot and plane had safely survived an acceleration of 11 g in the dive.

At the Reich Air Ministry too, the results were regarded as a sensation, and naturally the prompt utilization of such superior speed in this second year of the war was insisted on. But there was a great difficulty at this point, for the maximum altitude of the Me 163 A was about 6000 meters. But that was too low to attack enemy planes flying at 7000 to 9000 meters.

Now Hellmuth Walter had already developed the so-called "hot" powerplant, in which fuel was injected in addition to the decomposition of hydrogen. There were already takeoff rockets of this type with 1500 kp of thrust. This amounted — to express it in conventional horsepower figures — to some 4500 HP on the ground and up to 9000 HP at great heights. This powerplant was driven by two liquids, T- and C-fuel, uniting in the combustion chamber. T-fuel was about 80% hydrogen peroxide, while C-fuel consisted of 57% hydrazine hydrate, 30% methanol and 13% water. The catalysator was 0.6 g/l potassium-copper cyanide.

To be able to carry out the design and building of the Me 163 B combat rockets as quickly as possible, fifty engineers were added to the design office of Department L. On December 1, 1941 the first drawings were begun, and in April of 1942 the first finished cell left the factory to fly in tow behind an Me 110. During flying-in and subsequent flying at Augsburg and Rechlin, no essential

shortcomings in the cell were found, but an unpleasant delay occurred because the hot powerplant was ready for use only in the autumn of 1943, though work had gone on day and night at the Walter works in Kiel.

Test Command 16

In July and August of 1943, Test Command 16 of the Luftwaffe was established at Bad Zwischenahn in Oldenburg and given the task of carrying out combat testing of the Me 163 B. This was necessary because there was no previous experience with a plane that reached its combat altitude of 9000 meters in two to three minutes. In addition, the Walter powerplant, with about 2000 kilograms of fuel in its tanks, only ran for four to, at most, six minutes. After that the tanks were emptied except for a quantity of some 70 to 150 liters, after which only a glider landing was possible.

Above: The cockpit of the Me 163 B.

Below: That of the Ju 248 (Me 263).

Above (left to right): "Baron von Münchhausen", drawn by Hans Liska, was the symbol of Battle Squadron 400, which flew the Me 163 B in combat.
Right: The other emblems were those of echelons of JG 400.

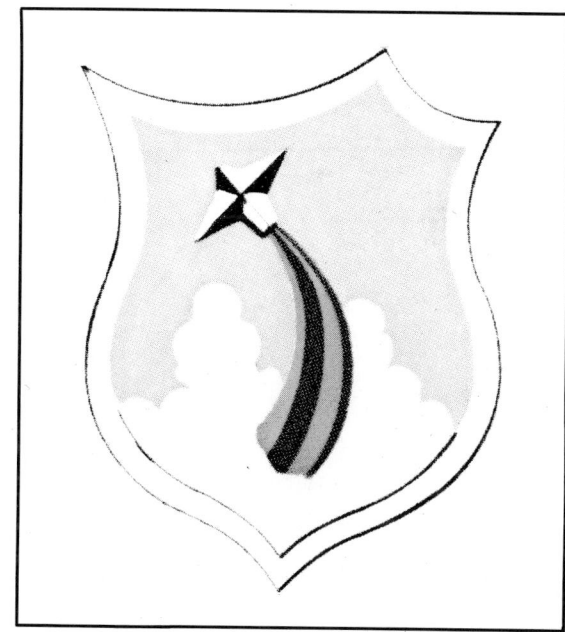

From both sides: An Me 163 B ready for takeoff at the test field in Bad Zwischenahn. It bears factory number 191916 and the squadron emblem of JG 400.

The takeoff was done with a two-wheel main landing gear plus a tail wheel. After liftoff, the main landing gear was dropped, and the skid and tail wheel were retracted at the same time. The skid and tail wheel were deployed again for landing. Takeoff and gliding according to calculations were impossible. Every landing had to be carried out as an unqualified target landing. If the plane came in too short or too far, this meant serious danger and usually the death of the pilot when the fuel remaining in the tanks exploded.

Upper right: Shortly after takeoff, the just-released main landing gear is still visible.

Lower right: An Me 163 B landing on its skid.

Below: Takeoff.

Because of the absolute necessity of ending every flight with a target landing, the pilots' first training was based on this condition. It began with normal glider training and ended with landing practice with the "Habicht" glider, whose wings were shortened to a breadth of 8 and 6 meters in order to attain higher landing speeds similar to those of the Me 163. After that, gliding flights with the Me 163 A began, and then with the Me 163 B, three versions of which were towed and landed during training, at first with empty tanks and no weapons, then with tanks filled with water, and finally with full tanks and weapons, corresponding to the later maximum takeoff weight. Between the towed flights with the 163 A and B, three to five live takeoffs were flown with the more harmless 163 A. Finally, six or seven live flights were made with the 163 B combat plane.

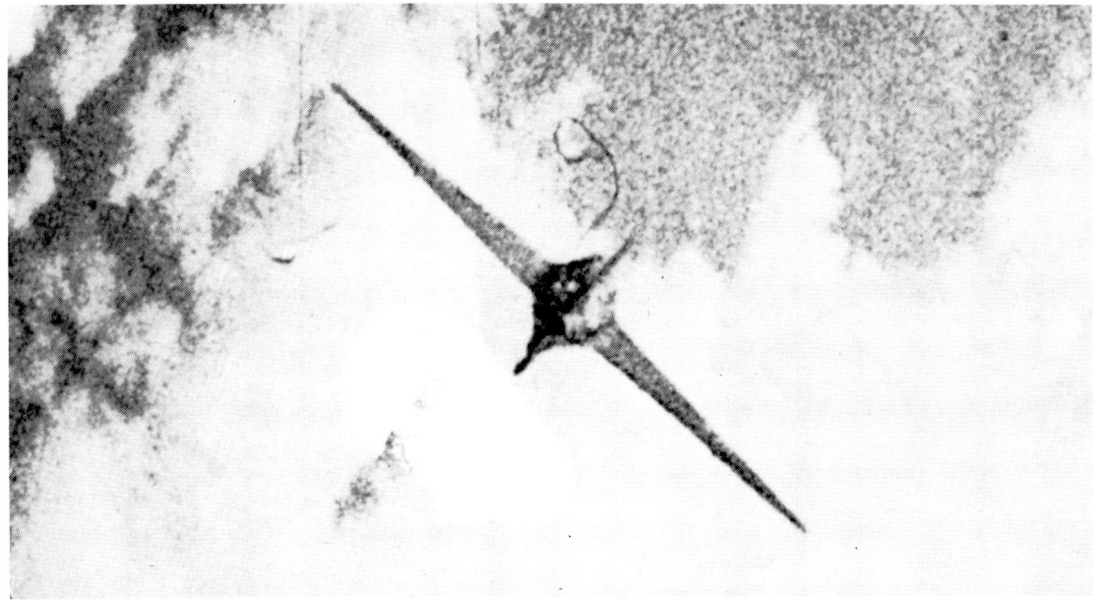

Opposite page, above: An Me 163 N, photographed from the towing Me 110, during combat testing at Bad Zwischenahn in Oldenburg. Below, an Me 163 coming down without landing gear after a mission. Far left: This also happened often enough in testing. The high explosive power of the C- and T-fuels in the tanks of the Me 163 B was very dangerous. A small leak sufficed to literally flatten the plane and its pilot. This series of pictures shows the phases of one of the serious, unfortunately frequent accidents during testing.

Above: The plane is moved out for takeoff.

THE FLYING CHARACTERISTICS OF THE ME 163

In three words, they were outstanding — if not unique. Thanks to its lack of a long fuselage with its control surfaces, the Me 163 was one of the most maneuverable airplanes ever built, if not the most maneuverable of them all. The lighter Me 163 A in particular could perform all maneuvers easily, as could the B, though it was a bit more sluggish on account of its greater weight. The A could even do a loop to the front when its tanks were empty. The only figure that could not be done was the spin. Neither the A nor the B could be made to spin, not even in the most advanced tests. If one pulled the stick all the way back during slow flight, it sank like a parachute, only faster. If one moved the stick to the left or right, the plane made a neat spiral downward.

In a landing flown according to the book, the target approach was no problem. Pilots soon grew accustomed to the higher approach speeds of ca. 180 kph in the A and 230 kph in the B.

The 163 took off from a paved runway, lifting off after it had rolled about 300 meters. After immediately jettisoning the heavy landing gear, by the time it reached the edge of the field, with an average-length runway of 1000 meters, its speed had reached 800 kph, at which speed it could ascend. Since the powerplant cut out promptly at negative acceleration, the pilot just had to make sure to push the stick forward strongly. It was best to end the ascending flight with a kind of upsweep and half roll, with no danger of negative acceleration.

The real dangers in this, the world's first rocket-driven combat plane, were not in its flying characteristics and only to some extent in the incomplete development of the first powerplant of its kind, but chiefly in the fuels, as concerned their highly explosive nature. If the slightest leak occurred anywhere that allowed the two fuels to make contact, an explosion was the result, and neither plane nor pilot could survive it. It happened several times that a pilot landed too fast, set down too late and slid off the runway on the skid. If he got onto rough terrain or ran into some inherently harmless obstruction on the ground, the plane could turn over. Even a harmless turnover could result in a leak in the fuel lines or tanks, and then it took only a fraction of a second until the machine exploded into a thousand pieces. There was no chance of survival. At least two planes exploded on the ground before takeoff, even before the powerplant was turned on. The pilots were killed.

If the pilots, who volunteered without exception for testing and combat duty in the Me 163 B, loved this plane better than any other, it was for two reasons: The fascination of the assignment, and the unparalleled experience of rocket flight in a plane with such outstanding flying characteristics.

Above: A plane taking off (at right beside the truck), at the start of which a cloud of white steam was given off.

Below: The plane has rolled forward and the steam had been blown backward.

Above: Preparing for combat: Before every takeoff the pilots put on their so-called PVC coveralls and boots, in order to protect themselves from being burned by the fuels, which sometimes happened. Below: How necessary this protective clothing was can be seen in this photo of Oberleutnant Franz Rösle, whose face was burned by leaking C-fuel. Fortunately, he suffered only first-degree burns, so that he could be treated and recovered.

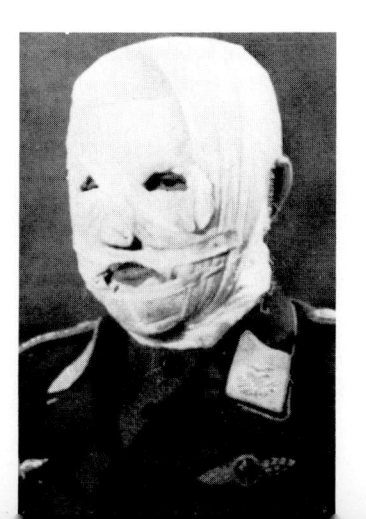

This Me 163 B exploded shortly after liftoff and lies on the ground in small and smaller pieces.
The rear fuselage and wings lie near the point of impact, totally destroyed.

Preparing to take off. Above: Feldwebel Ryll and his mechanic in a last cockpit check. Lower right: Feldwebel Schubert (with his inevitable unlit pipe in his mouth), and a thick armor plate in front of him. Both pilots died in action, in 1943 and 1944.

Upper right: One of the most successful German fighterpilots, Oberst Gordon M. Gollob, who succeeded Galland as General of the Fighter Pilots, visiting JG 300 in Brandis. Beside him is the Group Commander, Hauptmann Robert Olejnik.

Combat Testing and Combat

Flight testing consisted — aside, of course, from the training of the first pilots — of the following assignments: First of all, the physical testing of the pilots, the people who flew a rocket plane. Never before had people been exposed to such high speeds and acceleration, never before had a human being had to withstand a change in altitude of some ten kilometers in three minutes or less. Pressurized cabins did not exist at that time. Altitude sickness, inevitably ending in death, were a constant danger from any fault in the oxygen supply.

The quick withstanding of such great changes in altitude made a special diet necessary, consisting exclusively of non-flatulent foods. This was because all air held inside the body, even, for example, in sneezing, could lead to such raging pain in the head or body that one had to break off the flight and land as quickly as possible.

These biological tests were carried out by the specialist Dr. Duncker with the help of a large low-pressure cabinet captured in Russia. In it the pilots learned above all else to recognize the symptoms of altitude sickness promptly enough to take counter-measures before they lost the ability to control themselves.

Above: Hanna getting into the cockpit, and her portrait.

Hanna Reitsch, the only woman in the world to fly a rocket plane. She made numerous live takeoffs with the Me 163 A and towed takeoffs with the Me 163 B. She was not allowed to make live flights of the combat machine because there was simply too much concern for her.
Below: Two rare in-flight photos of an Me 163 B. They were taken by an American pilot during the war.

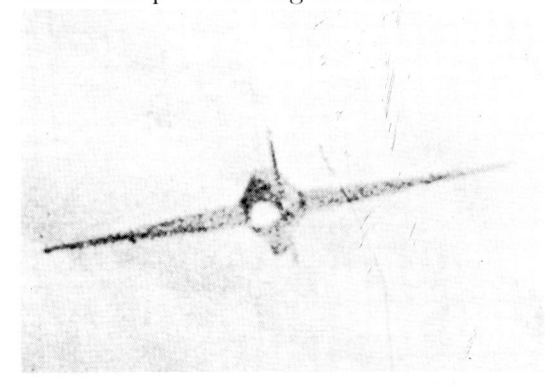

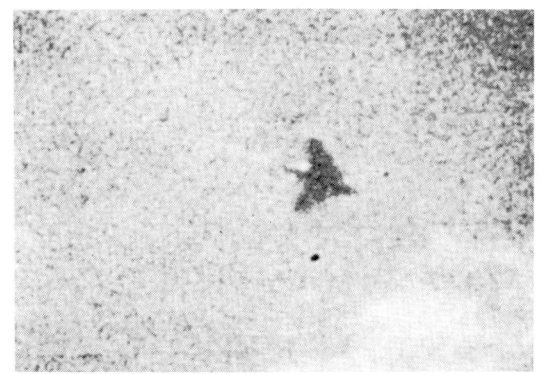

Here the divided powerplant of the Ju 248 — a parallel development to the Me 163 made by the Junkers works — is easy to recognize. Above is the main powerplant, below the cruising unit.

Practical combat training was necessary because the quick ascent of the Me 163 allowed completely different methods of attack. With no other plane was it possible to reach approaching bomber groups in two to three minutes. The Me 163 took off only when the bombers were in sight from the ground at the air base or reported in very close proximity. The attacking Me 163 B's had to reach their opponents in two to four minutes to be able to attack them effectively. This could be done either from below, behind or in front, using thrust, or in a glide from above when the tanks were empty. The duration of an attack rarely extended more than ten to fifteen minutes, since the powerplant ran for only four to six minutes at the most, as already noted. There were two types of attacks that seemed especially successful and were practiced. One was the "chicken ladder", in which the 163 was steered upward in a drawn-out zigzag course, the other the "spiral staircase" using a spiral ascending flight.

Controlling the Me 163 B in the air was done from the ground by using a Würzburg device. It functioned very well as a rule and also allowed flights and combat in less than perfect weather. The pilot could not fly the plane himself in complete cloud cover with a low ceiling, or in worse weather, as a long search for the landing field without power was impossible. If a pilot was not absolutely sure of reaching his base or — as happened a few times — another suitable field, then he had no recourse but to bail out.

40

While the Me 163 A and B took off on jettisoned landing gear and landed on deployable skids, their successor, the Me 263 (renamed Ju 248 by the manufacturer, Junkers) was to have retractable bow landing gear and an auxiliary cruising jet.

This Ju 248 — as well as a two-seater — was already in the prototype stage and was flown. But Germany's defeat put an end to its intended series production.

The great weaknesses of the Me 163 B showed up only in combat use. In its first offensive flights, the enemy crews of both fighters and bombers were very shocked by the vastly superior speed of the Me 163 B. But soon its "Achilles' heel" was discovered, and enemy fighters made good use of the knowledge. The Me 163 flying home in a glide after an attack was almost helplessly exposed to its attackers. In a few cases it was able to get away from an attacker by a dive from a high elevation, for even thus the Me 163 B reached speeds of 800 to 900 kph or was so maneuverable at lower speeds that the attacker could be shaken off, but at the end of its flight, at altitudes of some 1500 meters or less, it was no longer capable of any defensive moves, and at speeds of about 300 kph it was usually easy prey for fighters attacking from the rear. If they were near, they could calmly follow the Me 163 as it came in to land and fire on it. With only little success, attempts were made to protect landing Me 163's with strong anti-aircraft gun positions. Protection by conventional fighter planes such as the Me 109 or FW 190 rarely took place.

As of 1944 the Me 163 B was to be built and put into service by the Japanese. Only one place was built, though, and it was originally used for towing training. After its first and only flight with a powerplant, it was badly damaged while landing.

Above: A Japanese pilot stands before the Me 163 B built in Japan.

Below: The powerplant jet of an Me 163 B.

Test Command 16 of the Luftwaffe, established late in the summer of 1943 under the command of Knight's Cross holder Hauptmann Wolfgang Späte, originally consisted of five teachers and 23 selected, experienced fighter pilots. The teachers were Hauptmann Späte himself, Oberleutnant Rudolf Opitz, who ranked with Günther Groenhoff, Heini Dittmar and Späte among the pioneers of Me 163 flight testing, Oberleutnant Joschi Pöhs, who also held the Knight's Cross, Hauptmann Thaler and Oberleutnant Herbert Langer. This first group of rocket pilots was gradually expanded by additional volunteers.

The first combat echelon of Rocket Squadron JG 400, established early in 1944, was sent to Wittmundhafen on March 1, 1944, under Hauptmann Robert Olejnik, a holder of the Knight's Cross, with five planes and twelve pilots. The second echelon, under its captain, Hauptmann Otto Böhner, went into service at Venlo, Holland. A third echelon was organized later at Stargard under Hauptmann Opitz.

On account of the Allied invasion of Normandy that began on June 6, 1944, none of these three units saw combat action. The withdrawal from their bases that soon became necessary eventually united the whole squadron at Brandis, near Leipzig.

It was already too late for Echelons 13 and 14 of Replacement Fighter Squadron II, established late in the autumn of 1944 under Adolf Niemeyer and Mano Ziegler, who were able to train few students for lack of fuel. Both units were formed at the eastern Udet field, but had to retreat from advancing

TECHNICAL DATA

	Me 163 A	Me 163 B
Wingspan	8.85 m	9.30 m
Length	5.60 m	5.92 m
Wing surface	17.5 sq.m.	19.6 sq.m.
Dry weight	1450 kg	1900 kg
Flying weight	2400 kg	4300 kg
Powerplant	R II 203 b	HWK 509 A
	750 kg	1500 kp

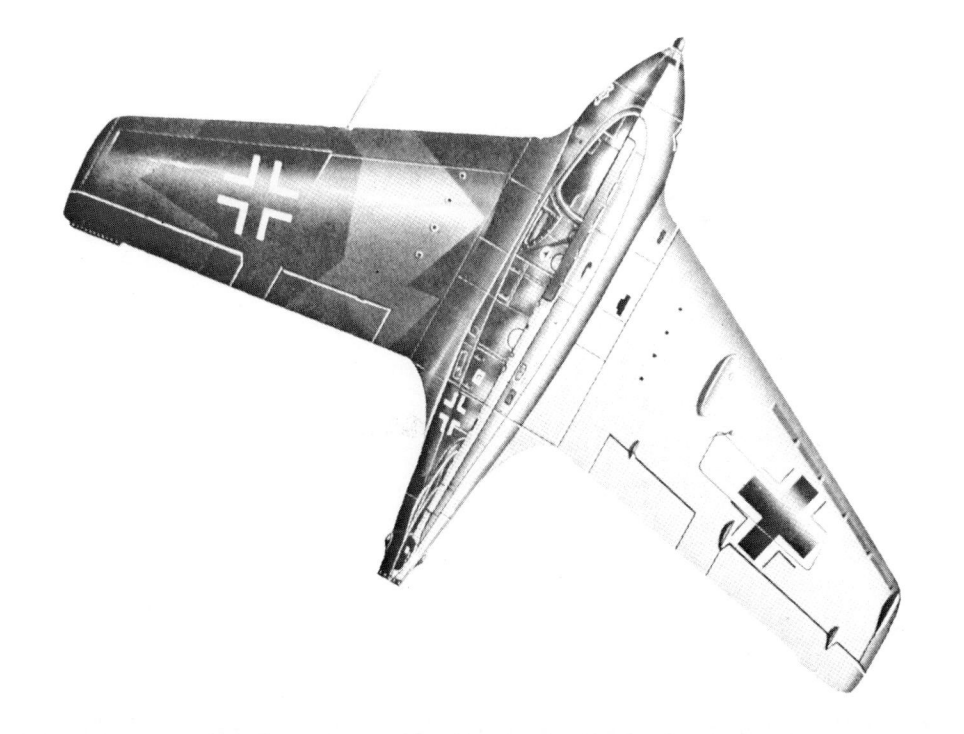

43

Soviet tanks in barely three weeks, after which they also gravitated to Brandis.

In Brandis the last act of this brief drama, which the fastest and presumably also the most unique fighter plane in the world could have performed, was played. At least it was the last act in a drama unlike any other in the adventurous history of aeronautics. The first and only Me 163 B rocket plane ever used in combat within a squadron unit showed its amazing, incredible superiority for the last time but at the same time took its last sacrifice. What was taking place in flight there was actually still testing, for the pilot could put only limited trust in his powerplant and was in danger of sudden explosions on the ground or in the air. But the excited mood of these pilots in combat was astonishing, and remained so despite the dangers. Thus in the last weeks of the war, Leutnant Hachtel and Leutnant Kelb tested vertical weapons ignited by selenium cells and consisting of five 5-cm guns built into each wing. They were ignited when the 163 B flew under an enemy plane. To test these weapons, a cloth about ten meters long and two meters wide was stretched between two wooden poles 20 meters high, and the planes flew under it. Hachtel and Kelb risked their necks and heads to fly between the two poles at almost 900 kph, but the cloth was torn to shreds, as was one or another bomber afterward.

Above: An Me 163 B captured by the British. It was displayed at the air show in Farnborough in 1946 and flown on tow.
Lower left: An American soldier guards an Me 163 captured at the Lechfeld air base.
Lower right: An Me 163 B-1 captured by the British in Schleswig-Holstein.

A captured Me 163 B-1 with factory number 191904. It is at the British airfield in Colerne today.

This captured FE 495 plane was displayed shortly after the war, along with other German and Japanese aircraft.

Measured in terms of cost, the success of JG 400 was limited, for on its last day of combat, April 16, 1945, the squadron could look back to having shot down only twelve enemy planes. And its essential and last assignment of protecting the Leuna works near Leipzig from further bomb attacks was unfulfillable. The enemy had air superiority.

The Me 163 has gone down in aeronautical history. Their development and use can be counted among those unusual events that pointed the way far into the future. Alexander Lippisch and Hellmuth Walter, along with the pilots of the Me 163, rank among those who took a first step into space. The next steps were taken by the Russians and the Americans, rewarded as victors with the experience and the co-operation of the vanquished. The hope remains that the end result will benefit our world.

Another captured plane. Below: The only "power egg" — as this plane was also nicknamed — in Germany. It stands in the German Museum in Munich today.

47

ALSO FROM:
Schiffer Military History

- THE WAFFEN-SS • THE HG PANZER DIVISION •
- THE 1ST SS ARMORED DIVISION •
- THE 12TH SS ARMORED DIVISION •

AND MORE...